D1200358

Unlocking the Secrets of Science

Profiling 20th Century Achievers in Science, Medicine, and Technology

Wallace Carothers and the Story of DuPont Nylon

Ann Graham Gaines

PO Box 619
Bear, Delaware 19701

Unlocking the Secrets of Science

Profiling 20th Century Achievers in Science, Medicine, and Technology

Marc Andreessen and the Development of the Web Browser
Frederick Banting and the Discovery of Insulin
Jonas Salk and the Polio Vaccine
Wallace Carothers and the Story of DuPont Nylon
Tim Berners-Lee and the Development of the World Wide Web
Robert A. Weinberg and the Story of the Oncogene
Alexander Fleming and the Story of Penicillin
Robert Goddard and the Liquid Rocket Engine
Oswald Avery and the Story of DNA
Edward Teller and the Development of the Hydrogen Bomb
Stephen Wozniak and the Story of Apple Computer
Barbara McClintock: Pioneering Geneticist
Wilhelm Roentgen and the Discovery of X-Rays
Gerhard Domagk and the Discovery of Sulfa Drugs
Willem Kolff and the Invention of the Dialysis Machine
Robert Jarvik and the First Artificial Heart
Chester Carlson and the Development of Xerography
Joseph E. Murray and the Story of the First Human Kidney Transplant
Stephen Trokel and the Development of Laser Eye Surgery
Edward Roberts and the Story of the Personal Computer
Godfrey Hounsfield and the Invention of CAT Scans
Christaan Barnard and the Story of the First Successful Heart Transplant
Selman Waksman and the Development of Streptomycin
Paul Ehrich and Modern Drug Development
Sally Ride: The Story of the First Female in Space
Luis Alvarez and the Development of the Bubble Chamber
Jacques-Yves Cousteau: His Story Under the Sea
Francis Crick and James Watson: Pioneers in DNA Research
Raymond Damadian and the Development of the Open MRI

Unlocking the Secrets of Science
Profiling 20th Century Achievers in Science, Medicine, and Technology

Wallace Carothers and the Story of DuPont Nylon

First Printing

Library of Congress Cataloging-in-Publication Data
Gaines, Ann.
Wallace Carothers and the story of DuPont nylon / Ann Gaines.
p. cm. — (Unlocking the secrets of science)
Includes bibliographical references and index.
Summary: Examines the life and work of Wallace Carothers, who invented nylon for the DuPont Company.
ISBN 1-58415-097-1
1. Carothers, Wallace Hume, 1896-1937—Juvenile literature. 2. Nylon—Juvenile literature. 3. Chemists—United States—Biography—Juvenile literature. [1. Carothers, Wallace Hume, 1896-1937. 2. Chemists. 3. Nylon.] I. Series.
QD22.C35 G35 2001
547'.0092—dc21
[B] 2001029994

ABOUT THE AUTHOR: Ann Gaines holds graduate degrees in American Civilization and Library and Information Science from the University of Texas at Austin. She has been a freelance writer for 18 years, specializing in nonfiction for children. Some of her recent books include *Sammy Sosa* (Chelsea House) and *Britney Spears* (Mitchell Lane). She lives near Gonzales, Texas with her husband and their four children.

PHOTO CREDITS: pp. 6, 14, 34, 40 Hagley Museum; pp. 22, 27 Superstock; p. 28 Archive Photos

PUBLISHER'S NOTE: In selecting those persons to be profiled in this series, we first attempted to identify the most notable accomplishments of the 20th century in science, medicine, and technology. When we were done, we noted a serious deficiency in the inclusion of women. For the greater part of the 20th century science, medicine, and technology were male-dominated fields. In many cases, the contributions of women went unrecognized. Women have tried for years to be included in these areas, and in many cases, women worked side by side with men who took credit for their ideas and discoveries. Even as we move forward into the 21st century, we find women still sadly underrepresented. It is not an oversight, therefore, that we profiled mostly male achievers. Information simply does not exist to include a fair selection of women.

Contents

In the 1940s, women started to buy a product we take for granted today: nylon stockings. This ad by the DuPont company, which developed nylon, displays the great interest shown in these stockings, which replaced much more expensive silk stockings.

Chapter 1

An Amazing Product: Artificial Silk

In September, 1945, the battles of World War II finally came to an end when Japan surrendered. But almost immediately afterward, battles of a different sort broke out in U.S. department stores. Newspaper headlines said things like "Women Risk Life and Limb in Bitter Battle." Thousands of women would wait in line for hours, then, as one newspaper in Pittsburgh reported, "A good old fashioned hair-pulling, face-scratching fight broke out in the line." The winners would happily go home with their prize: nylon stockings.

What was the cause of these "nylon riots?" Why were women in this country so desperate to get something that today we take almost for granted?

The explanation is a fascinating story that involves an ancient Chinese princess and a lonely chemistry genius who lived more than 4000 years apart. It also involves tiny insects and huge industrial plants. Fabrics that at first were reserved for royalty and now are commonly used by almost everyone are at the heart of this amazing tale.

The story begins about 2600 B.C. According to Chinese tradition, Hsi ling shi, the 14-year-old wife of Emperor Huang Ti, accidentally dropped a cocoon of a certain kind of caterpillar into her tea. The cocoon unraveled before her eyes, revealing a long thread.

She did the same thing with several other cocoons and wound the individual filaments together to form a thicker thread. With the help of her ladies-in-waiting, she obtained many cocoons and used the threads to make a cloth that was very smooth and soft, yet also very strong. This cloth could be dyed many different colors and Hsi ling shi made a beautiful new robe for her husband.

Silk had been discovered.

The Chinese soon learned that silk came from a blind, flightless species of moth called Bombyx mori that lived for only a few days after emerging from its cocoon. But during their short lives the female moths would lay 500 or more tiny eggs at a time. When they hatched, the newborn caterpillars—which were known as silkworms—had huge appetites. But they would eat only a certain kind of mulberry leaf.

They would eat almost constantly for about a month, and with thousands and thousands of these little creatures all chomping away at once, the sound was like a heavy rain falling on the roof. They ate up to 50,000 times their original body weight and became very fat. During this time, they needed to be kept warm and away from loud noises and other things like human sweat.

Then they would prepare themselves for their transformation into moths by creating cocoons. Silkworms have a pair of salivary glands called silk glands which produce a clear fluid that is forced through openings in their mouths. It hardens as it comes into contact with the air and forms the filaments that the moths would spin into their cocoons. When the cocoons were complete, the caterpillars would become dormant and enter the pupal stage.

But when humans were raising silkworms, they would not let most of them become moths. If the pupa grew to become a moth and then emerged from the cocoon, that would damage the threads so they couldn't be used for the creation of silk. Therefore the cocoons would be baked or steamed to kill the pupa inside, then placed in water and unraveled. Only a few were kept alive so that they could produce more eggs and keep the production cycle going.

Sericulture, as the process of silk production was known, is very time-consuming. Because it takes the leaves of two full-grown mulberry trees and 4,000 silkworm eggs to produce one pound of reeled silk, it's easy to understand why silk became one of the world's most valuable and expensive fabrics. For thousands of years, silk clothing was the mark of an important person. It was known as the "queen of fabrics."

The Chinese kept much of it for themselves, but also sold some to other countries. They would load it onto caravans and send it over what was known as the Silk Road, a track thousands of miles long that wound through harsh deserts, barren plains and cold, windswept mountains.

But there was one thing the Chinese would not send. They kept how silk was made a secret and put to death anyone who tried to reveal that secret. No one outside of China knew that it was the product of a humble moth, and many wild guesses were made.

But it is impossible to keep something a secret forever, and not long after the birth of Christ other countries finally discovered how to make silk. According to one story, it came to Europe when two monks hid silkworm cocoons in their hollowed-out walking staffs. The secret also spread to Japan and Korea, and those two countries soon became among the world leaders in the production of silk.

But one thing remained constant for centuries after the Chinese lost their monopoly on silk: it was still reserved for nobility. Many European countries passed what were called "sumptuary laws," which forbade the wearing of silk by commoners.

Eventually more and more people began wearing silk garments, though it always maintained its reputation for fineness. In New York City, for example, the fashionable "Upper East Side" of Manhattan is known as the "silk stocking district."

During the 1930s, the U.S. imported most of the silk that was produced in the world. And nearly all of that imported silk was used to manufacture women's stockings.

But there were problems. One was that the price of silk could go up and down wildly. Another was that silk had to be imported, which left the US at the mercy of foreign countries.

And at that time there was one foreign country that the US was becoming increasingly alarmed about. It was also the country that produced most of the silk that the US imported. That country was Japan.

In the early 1930s, Japan invaded China and the US began putting pressure on the Japanese to withdraw.

The Japanese refused and relations between the two nations began to worsen.

One young woman wrote: "I do not intend to pay for a single silk stocking until the Japanese get out of China."

Then on the morning of October 21, 1938, a startling headline appeared on the front page of the New York *Times* newspaper: "$10,000,000 Plant to Make Synthetic Yarn; Major Blow to Japan's Silk Trade Seen."

The story went on to say that two new manufacturing plants were being constructed by the DuPont Chemical Company. They were going to make a new man-made silk.

This new "silk" would be called nylon, and it was first made in a laboratory, not in the stomach of an insect. Charles M. Stine, the DuPont official who announced the creation of nylon to the public, noted that "I am making the first announcement of a brand new chemical textile fiber. This textile fiber is the first man-made organic textile fiber prepared wholly from raw materials from the mineral kingdom."

By the time nylon was introduced for sale to the public, DuPont had spent an enormous amount of time and money on it. Its invention had taken 11 years of research, cost $27 million, and involved more than 230 scientists and technicians.

The commercial production of nylon began in December 1939. In May 1940, when four million pairs of nylon stockings—the first ever—went on sale, they sold out within a few hours.

But when the Japanese bombed Pearl Harbor, a naval base in Hawaii, the U.S. entered World War II.

By that time, nylon had become so important and had so many uses that almost all of it that could be produced went to the war effort. Nylon stockings, or "nylons" as they were quickly called, were almost impossible to find. Stockings that had cost $1.50 before the war were now selling for $12.00 and more—if people could find them. Some women became so desperate that they painted "stockings" on their legs with brown makeup and used eyebrow pencils to draw "seams."

Even when the war was over, it took DuPont nearly a year to produce enough nylon to meet the demand for women's stockings. Then the "Nylon riots" finally ended.

Over the years, factories started to make more and more items out of nylon. By 1980, seven and a half billion pounds of the material were sold. Today, the DuPont Company sells nearly five billion dollars of nylon every year.

Nylon was created as the result of the work of a team of organic chemists at the DuPont Company that was led and directed by Wallace Hume Carothers, one of the greatest chemists of the twentieth century. How Carothers and his team found this useful product almost by accident is a fascinating story.

You might think that someone who was responsible for such an important product would have been a very happy man. But that was not the case with Wallace Carothers. He was unhappy for much of his life.

While no one knows the exact reasons, this young man killed himself before nylon even went into production. He never got to see how important his invention became, nor how many people benefited from his work.

Wallace Hume Carothers was one of the greatest organic chemists of the twentieth century. He was working for the DuPont Chemical Company when his team of chemists created nylon.

Chapter 2

Early Life

Wallace Hume Carothers, the inventor of nylon, came from a family that had lived in America since before the Revolutionary War. His ancestors moved from their original home in Pennsylvania to the Midwest and Wallace was born on April 27, 1896, in Burlington, Iowa.

Iowa was famous for the number of religious groups that moved to its flat farmlands to practice their religion in peace. Cedar Rapids in central Iowa was the home of the German Amana Community, a famous commune that existed from 1855 to 1932. Iowa City was the departure point for the Mormon expedition to Utah in 1856.

Like many of his neighbors, Ira Carothers, Wallace's father, was a deeply religious man. He was an elder in the Presbyterian Church in Burlington. A trim, small man who had lost most of the hair on the top of his head at an early age, he wore glasses and sported a large bushy mustache. He was very serious in nature.

Wallace was an obedient son and always showed his father great respect. But Wallace had a hard time being friends with his dad. All of their lives they had problems talking to each other. Wallace was much closer to his mother, Mary.

By 1901, Wallace's family moved further west in Iowa to Des Moines, its largest city. By then, the family had grown to include another son, John, and a daughter, Isobel. Wallace would eventually be the oldest of four children.

Ira began a new career in Des Moines. He served as vice president of Capital Cities Commercial College, a school that provided a one-year course in the basic skills of business like bookkeeping, letter writing, and business math. Students came to the school so they could get jobs in the business world.

Wallace was a happy little boy in the family home on Arlington Street in Des Moines. The house had a barn behind it and the Carothers kids had a large open yard in which to play. The kids played mostly with themselves. They became very close to each other.

Years later, his dad described Wallace in Matthew Hermes' book *Enough for One Lifetime.* "As a growing boy," Ira said, "he had a zest for work as well as play. He enjoyed tools and mechanical things, and spent much of his time in experimenting. He was a great admirer of Thomas A. Edison and seriously considered electrical engineering as a profession."

He added that Wallace liked reading. In an age before television or the movies, books were a major source of amusement for children.

"Very early in life he displayed a love of books. From the time when *Gulliver's Travels* would interest a boy, on through Mark Twain's books, *Life of Edison*, and on up to the masters of English Literature, he was a great reader," his father said in Hermes' book.

Wallace also fell in love with science at an extremely early age, almost as soon as he could read. A quiet, studious boy, he followed the directions he found in popular magazines and books to make electric door bells and radio sets using wire and Quaker Oats cereal boxes. He and several of his friends—who quickly nicknamed him "Doc" and "Prof"—formed a club to conduct electrical experiments.

Always a good student, he entered North Des Moines High School in 1910. As a teenager, he worked for the city reading water meters and at the public library as a clerk to make some spending money. He spent much of it at the hardware store buying wire, batteries, and other supplies for experiments he conducted in his bedroom, which he had converted into a laboratory.

Eventually he developed a special love for chemistry. He would remember in Hermes' book that "My interest in chemistry was started by reading Robert Kennedy Duncan's popular books while a high school student in Des Moines, Iowa, so that after some delay when it was possible for me to go to college I had definitely decided to specialize in chemistry."

The author he spoke of, Robert Kennedy Duncan, was an industrial chemist and professor of chemistry at the University of Kansas who wrote a series of books for young readers that explained the new and exciting scientific work being done in chemistry. In 1905, his book *The New Knowledge* was published. It was followed in 1907 by *The Chemistry of Commerce*, another of his popular readers.

In these books Duncan described the wonderful opportunities that existed for chemists to use their growing knowledge to create wonderful new substances that could make them rich. Thousands of young readers around the country such as Wallace Carothers read Duncan's books and others like them and were inspired to create more inventions.

Because Wallace was so interested in science and chemistry, it seems a little surprising that he enrolled at his father's commercial college in Des Moines and took classes in shorthand and bookkeeping after he graduated from high school in 1914. Classes like that would have had very little value for someone who aspired to a career in science. Some people believe that Wallace, a shy boy, did what his father wanted him to do rather than sticking up for himself.

When he graduated the following year, he attended Tarkio College, a small Presbyterian Church college located in northwestern Missouri. His father worked out an agreement with the college administrators so Wallace could attend. Wallace would teach Tarkio's business courses part-time in exchange for his tuition and other expenses.

Wallace Carothers was 19 years old when he began at Tarkio, a year older than most of the new freshmen. His classmates were the sons and daughters of Presbyterians who lived in the area. At Tarkio, classes were small. In 1915, there were only 58 men and 56 women enrolled at the school. Everyone knew everyone else.

Every student took Greek, Bible, and Ethics. Many of the students planned, after school, to be missionaries and ministers. Everyone was required to attend chapel every day and they sang psalms at services. Wallace had a beautiful voice and enjoyed singing very much.

At Tarkio, he did well in his basic classes. He also did extremely well in his German classes, becoming fluent in the language. German was then the language in which important research in the field of chemistry was published. Knowing the language gave Wallace an advantage in studying the latest discoveries in his favorite subject.

But the best thing about Tarkio was that Wallace finally had the chance to learn some real chemistry. He studied in the school's basement laboratory under the direction of Arthur Pardee. Professor Pardee had recently completed his doctoral dissertation and was up-to-date on all the latest research. Wallace wasted no time: by the end of his third year at Tarkio, he had already taken all the chemistry courses offered at the school. By this time, he knew what he really liked. His favorite type of chemistry was organic chemistry, the chemistry of life.

He learned that during the nineteenth century, chemists began to separate complex substances found in nature into their smallest parts, which were called elements. The elements of carbon, oxygen, hydrogen, nitrogen, and a few others make up almost all living substances. Carbon, the most plentiful of these elements, is found in all living things. Pure carbon is what makes up charcoal or the black particles in the soot of a candle.

All elements, including carbon, are atoms, which are incredibly tiny. The atoms of different elements can be combined to form compounds. For example, the elements hydrogen and oxygen form water. The smallest particle of a compound is called a molecule, and it consists of atoms of the elements that compose it. A water molecule contains two atoms of hydrogen and one atom of oxygen.

One of the greatest discoveries of the time occurred when chemists found out that the carbon elements in a molecule of an organic compound have the ability to link to themselves in chains, and several of these different carbon chains can link together. They figured out that it is possible to have an organic substance with ten or fifty or even a thousand carbon atoms linked to each other in several different chains.

Because Wallace understood concepts such as these so well, the administrators at Tarkio asked him to become the chemistry instructor when Arthur Pardee left Tarkio to go teach at another school in 1918. Wallace postponed his graduation and taught chemistry at Tarkio for two years. During that time, the basement laboratory was his playground. Once the students had gone for the day, he could use the equipment for his own experiments.

Wallace constructed elaborate distillation towers to capture liquids. In test tubes by the hundreds he isolated intensely colored liquids that slowly solidified into beautiful natural crystals more intense and beautiful than any polished jewels.

Once a pure sample of the new compound was isolated in a test tube, it was destroyed there, usually by burning. What was left in the test tube he then studied to determine which elements, and how much of each, were contained in the molecules of the new compound. Results from these experiments were obtained using sealed analytical balances of the greatest delicacy and cost.

Wallace became a remarkably good organic chemist during this period. His experiments were clean and precise. His notes were complete, his preparation was thorough. He earned a degree from the tiny college in 1920 at the age of 24. He had made no grade lower than an A all of the time he was at Tarkio.

In college, Wallace Hume Carothers took all the chemistry courses he could. He learned from his professor's lectures. But what he really liked was doing experiments in the laboratory.

Chapter 3

Education

In 1920, just as today, an undergraduate degree in chemistry was simply the first step in the formal education of a chemist. When Wallace Carothers received his undergraduate degree in chemistry from Tarkio College, he applied to the Graduate School of Chemistry at the University of Illinois. He was awarded a graduate assistantship from the Chemistry Department.

This meant he received free tuition and would be paid a small wage to teach the basic chemistry laboratory course to undergraduate students. Chemistry professors were not expected to teach laboratory techniques to new students. Professors taught advanced students. Incoming freshmen were usually taught by graduate students. Throughout most of an ordinary day, these graduate students would supervise the experiments of hundreds of younger students. They would hand out the proper chemicals and guard against dangerous situations. Afterward, the graduate students were responsible for cleaning the dirty test tubes and flasks. Usually it was boring work, doing the same basic steps over and over.

But in the evenings and on weekends, these graduate students retreated to a small laboratory space of their own where they would study and conduct their own experiments. Wallace quickly established a reputation among the faculty at Illinois as one of the most capable graduate students and within a year, he had received his first graduate degree, a Master of Science.

Because he didn't have much money, he accepted a teaching position at the University of South Dakota, where his old professor from Tarkio, Arthur Pardee, was chemistry department chairman. But the young chemist, probably because of his shyness, was not considered to be an especially good teacher. So he returned to the University of Illinois after just one year, where he was happier working in the chemistry laboratory.

Wallace became known to the other chemists at Illinois—both students and faculty members—for his bright, comprehensive, and inventive mind. So it wasn't surprising that the chemistry department gave him a job as an assistant in February, 1924. This was both a giant step and a baby step at the same time in Wallace's career and in his mind. He was employed full-time in the Chemistry Department. He stopped receiving his scholarship of $75 a month. Instead he would be paid a salary of $1800 a year. It was just a little more money, but Wallace was now recognized by other chemists as a brother.

Three months later, Wallace received word that he would be awarded his Doctor of Philosophy—or Ph.D—degree. The Ph.D. is the highest academic honor awarded in the United States.

Carothers remained at the University of Illinois for two more years as an assistant. He taught classes and supervised advanced laboratory courses. It was a good time for Wallace. He had four or five friends among his students and colleagues who all loved chemistry. The hours they spent in study and lab were fun times, hardly any work at all.

But he did not stay there. Harvard University in Cambridge, Massachusetts, one of the nation's oldest and most famous universities, offered Wallace a job. Early in 1926, its chemistry department asked him to teach organic chemistry there, starting in the fall. They offered him a salary of $2250. Carothers accepted.

But he did not go directly from Illinois to Cambridge. Instead, before he moved into new quarters at Harvard, Wallace went to Paris with his friend and co-worker, Jack Johnson, for a well-earned vacation. They left from New York on June 13, 1926 on a French ocean liner named the *Paris*.

In early September, Wallace, Jack, and a French chemist friend, Gerard Berchet, boarded the *Paris* for the return trip to the United States. They had purchased six bottles of champagne in France, which they brought aboard. They drank one bottle each night, knowing they could no longer drink when they reached New York. This was because it was the time of Prohibition, a period in American history when it was illegal to buy or sell any alcoholic beverages.

Settling down after his visit to Europe, Wallace taught three semesters at Harvard. He published one scientific paper. But he was unhappy at Harvard because his teaching duties kept him away from experimenting with projects that interested him. It was hard to find the time and money he needed for original research. And how he ached to do research. At Harvard, he had become interested in one particular problem: polymers. Polymers are chemical compounds made up of many smaller repeating molecules.

Rubber, wool and silk are examples of polymers that occur naturally.

Although many chemists had tried to make large polymers similar to natural ones in the laboratory, no one had been able to do it. Wallace became really interested in these polymers and wanted to spend all of his time doing this kind of research.

By this time he had also realized that he did not like teaching, nor was he very good at it. Although he liked talking with his friends, he was still a very shy young man who found it uncomfortable to be in front of groups of people.

Therefore, he did not want to stand behind a podium and lecture. He was much happier alone in the lab. Instead of teaching, he wanted to do pure research. Luckily, there was a large chemical company that was looking to hire just the kind of chemist Wallace wanted to become.

By the time he was 30, Wallace Hume Carothers knew that he was happiest when he was alone in a chemistry laboratory. He had become very interested in creating polymers.

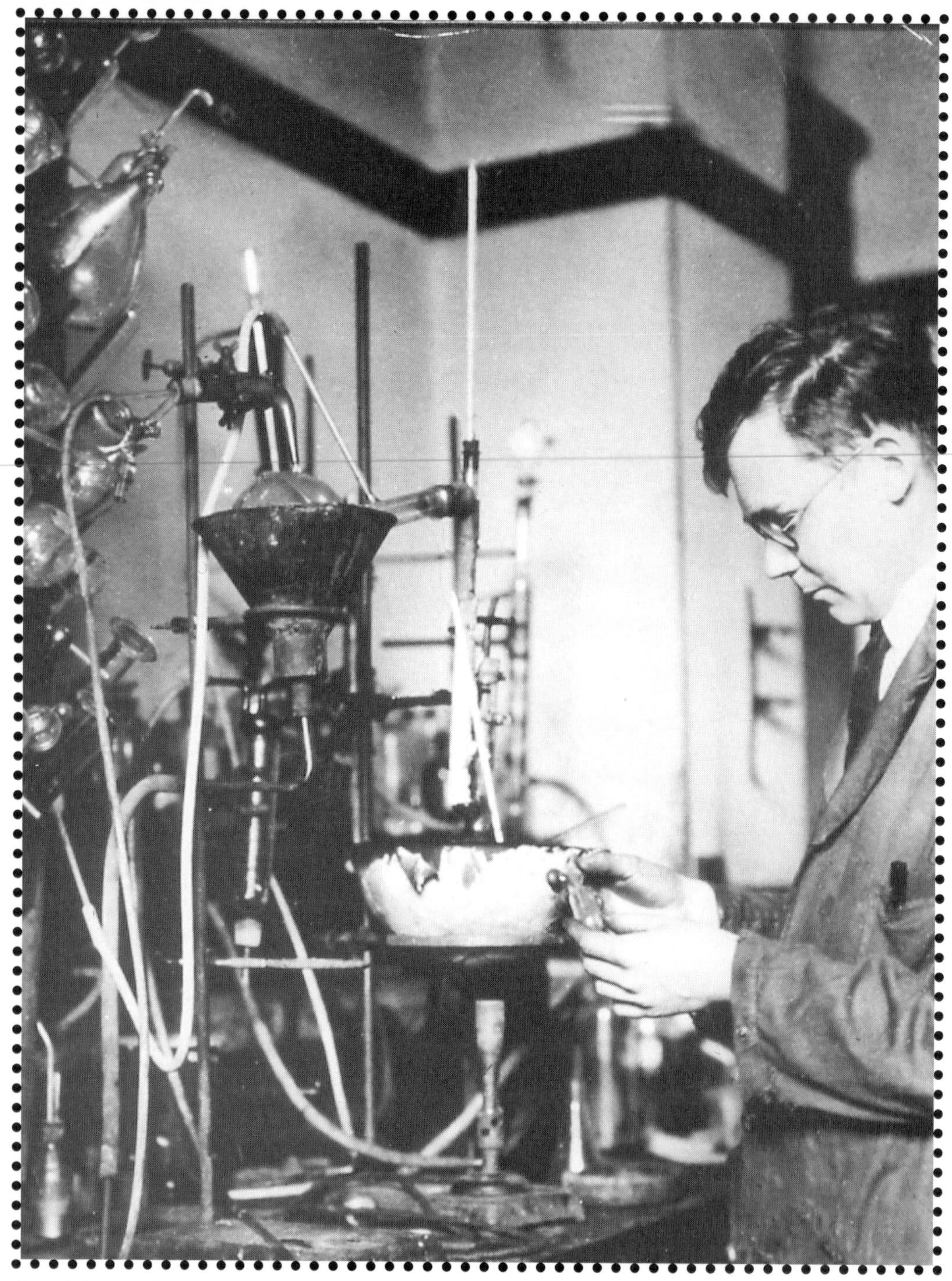

In 1928, Wallace Hume Carothers went to work for the DuPont Chemical Company. The company hired him to undertake new research. This photograph was taken of Carothers in his laboratory at DuPont.

Chapter 4

The DuPont Company

The DuPont Company was one of the oldest continuously operating industrial enterprises in the world. The company was established in 1802 near Wilmington, Delaware, by a French immigrant, Eleuthère Irénée du Pont de Nemours, to produce black powder or gunpowder. E. I. du Pont had been a student of Antoine Lavoisier, often referred to as the father of modern chemistry, and he brought to America some new ideas about the manufacture of consistently reliable gun and blasting powder. The DuPont company made a huge fortune selling gunpowder to the Union during the American Civil War and an even greater fortune selling more than a billion pounds of guncotton, another explosive, to the United States Navy during World War I.

Charles Stine was the man who had assembled the first explosives that DuPont sold to the United States Navy at the beginning of World War I. He was appointed the director of the Chemical Division of DuPont in 1924. Three years later he wrote to the directors of the company that "We are including in the budget for 1927 an item of $20,000 to cover what may be called, for want of a better name, pure science or fundamental research work...the sort of work we refer to...has the object of establishing or discovering new scientific facts."

The following year, Stine was able to persuade the directors of the company to set aside $250,000 to fund research into subjects that had no immediate chance to produce profitable products for the company.

They even put the project into its own building, which they called the DuPont Experimental Station. But most people at DuPont jokingly called it "Purity Hall" because it was pure, or basic, research.

At first, DuPont couldn't find anyone to become director of the organic chemistry division. Famous scientists who worked at colleges and universities felt that working in a "factory" was not dignified enough for them.

But Dr. Roger Adams of the University of Illinois, one scientist who had turned down DuPont, recommended that they interview Carothers. Wallace had been a former student of his and Adams had been very impressed with him.

In September 1927, Wallace Carothers interviewed with Stine for a job in their new pure research division. Even though it was the kind of job that Wallace was looking for, he was reluctant to take it. Because he had such a hard time making friends and being around people, he was afraid of going some place new where he didn't know anyone.

So his first reply was "no."

But the DuPont Company recognized what a brilliant chemical mind Wallace had. They continued to pursue him and finally made him an offer of a salary nearly twice what he was making at Harvard. They were more than willing to trust in his genius to eventually produce valuable new products from his research. In addition, they promised that he could set his own research goals.

So he reported for work at the DuPont Experimental Station in February, 1928.

Right after he was hired, he wrote to John Johnson, the friend with whom he had traveled to Europe, that "A week of industrial slavery has already elapsed without breaking my proud spirit. Already I am so accustomed to the shackles that I scarcely notice them."

At first, Wallace spent most of his days alone, reading in the DuPont Chemistry Library. No one asked him what he was doing or when he would have some new results for them to inspect. It was a wonderful time for Wallace. He had the time and money to follow his imagination.

He later told a friend, "As for funds, the sky is the limit. I can spend as much as I please. Nobody asks any questions as to how I am spending my time or what my plans are for the future. Apparently, it is all up to me."

Sometimes he would spend hour after hour in his laboratory, working on a project that interested him. To relax, he enjoyed having one or two friends over in the evening to listen to classical music. He sometimes said that if he hadn't gone into chemistry, he would have liked to become a classical musician.

And even though he was a very serious man, he also had a sense of humor. One time when he hosted a party for some of his co-workers, he served them what he called "rare Pacific Ocean mollusks." But they turned out to be small chunks of synthetic sponges that were manufactured by another DuPont department and the surprised guests quickly spit them out.

In the first years after he joined DuPont, Wallace Carothers hired a small team of outstanding research chemists to work with him. Many of them had been recommended by his old friends at Illinois and Harvard. In 1929, Wallace and his team spent $43,000 on their research. In his report that year, Charles Stine said that Carothers and his team were studying polymers but nothing had come of it yet.

But that changed very soon. By March 1930, Carothers' team could do an amazing number of things with their new polymers. They could make polymer chains of organic compounds that also contained the element silicon. DuPont would eventually use this idea in car finishes and lubricants.

And in April that year, Wallace had his team experiment with a polymer known as vinyl acetate. They studied what happened to it when it reacted with hydrochloric acid. They thought at first that the experiment was a failure because the expected results were not obtained.

But Arnold Collins, a member of Carothers' research team, noted a small white solid that had gathered as a residue of the experiment that had been done in one of the flasks waiting to be washed and cleaned. He fished the material from the inside of the flask with a long thin wire hook. It felt like rubber and bounced when Collins showed it to Julian Hill, another worker in the lab. This was the first synthetic rubber.

Carothers immediately saw the profound importance of this discovery. He formed a special group to investigate this important new chemical and devise a safe and effective way to produce it in large quantities.

It was truly just an accident that Arnold Collins had been curious enough to investigate the trash of an experiment. This type of scientific discovery is called serendipity. It is unexpected and unplanned for. But DuPont took full advantage of the serendipitous find. The company called the new substance Duprene. Soon they changed the name to neoprene.

Neoprene is considered superior to rubber in many ways, such as its resistance to sunlight, abrasion, and temperature extremes. These properties made it popular in many industries. For instance, neoprene is favored for electrical cable insulation, telephone wiring, and roofing. It is also used for wetsuits, which many people wear when they go scuba diving or surfboarding.

In this photograph, Wallace Hume Carothers shows off neoprene, or artificial rubber. His team of organic chemists created neoprene by accident, but it has proven extremely useful over the years.

Chapter 5

The Story of Nylon

Wallace Carothers considered Neoprene a lucky bit of science, but its creation had little to do with his primary interest, the creation of long carbon-chained polymers. So while part of his team worked on Neoprene, he spent most of his time with polymers. The Carothers research team believed that the reason that they could not produce any really long polymer chains was because their experimental method left very small traces of water in the reaction flask. This water was produced as a byproduct of the formation of the polymer chains in the reaction mixture, and these small traces of water reacted with the carbon chains and prevented larger molecules from forming. So Carothers decided to drive all of the water from a cooling mixture of the polymer by using a newly invented machine, the molecular still.

A molecular still worked at very low air pressures. This low pressure forced any water molecules that were formed in the boiling experimental mixture to leave the mixture and fly into the almost perfect vacuum above it. Once the water molecules left the experimental mixture, they were immediately frozen into a thin layer of frost on a cold metal rod suspended above the mixture. This kept the atmosphere above the experiment free of water vapor and ready to remove any more water molecules that were generated by the forming polymer chains in the reaction liquid.

Once the still was ready, Julian W. Hill—one of the chemists who worked with Carothers—put together another batch of a polymer they'd been studying. Hill moved the product to their molecular still and turned on the pump. Soon there was almost a perfect vacuum inside the reaction vessel. He turned on a gentle heat and warmed the waxy solid to about 200 degrees celsius. It soon melted and boiled in the near vacuum above it.

After two days, about 5% of the weight of the original waxy material was frozen as tiny ice crystals on the cold metal condenser above the warm liquid polymer that still remained in the reaction vessel. For the next five days, no new ice formed on the condenser. There was no more water left in the polymer below to prevent the formation of very long chains of molecules. On the morning of April 30, 1930, Hill turned off the still, disassembled it, and examined the tough polymer mass that clung to the insides of the reaction vessel.

The product had become a waxy solid. The team quickly figured that the prolonged heating without the presence of water had fundamentally changed the structure of the original polymer and allowed it to form even larger polymer molecules. It was a great success for Carothers, but he had achieved much more than even he imagined.

Hill heated the reaction vessel and then touched a glass rod to the sticky goo at the bottom of the flask. As Hill moved the rod away from the polymer mass, long thin silk-like fibers stretched from the glass rod back to the surface of the polymer and could be pulled from the flask and stretched all over the laboratory. The fibers showed great tensile strength. They were elastic, transparent, and shiny, very similar to natural fibers such as silk.

But these first fibers were not very strong, and they melted at too low a temperature. Under hot water, the fibers simply melted back into the sticky ooze of the original polymer. These new organic fibers opened the scientist's eyes to a new and marvelous discovery. But it was just a beginning. If it was to prove a useful and profitable product, it had to be stronger and be able to withstand harsh physical conditions.

All of a sudden, the businessmen at the DuPont Company recognized the importance of Carothers' work. A gooey polymer was worth nothing, but a strong artificial fiber with the properties of silk was worth a fortune. No longer would the silk market depend upon the output of millions of tiny caterpillars. Artificial silk could be produced in a factory by the ton. DuPont gave Carothers more money and more chemists. He was instructed to stop his original research into the nature of polymers and to search for the best artificial fiber.

For three years, from 1930 to 1933, Carothers and his team searched for the right kind of polymers to make commercial strength fibers. From hundreds of starting organic materials the team began to produce polymers. They tested each one for the right properties that would be needed for a man-made fiber to take the place of expensive imported silk. The correct polymer also had to be cheap to produce, using materials that were both low in cost and readily available to the scientists.

Early in 1934, Carothers decided to investigate a different family of organic compounds for the starting materials for the artificial silk they envisioned. Until that time, Carothers and his team had investigated polymers which were only composed of carbon, hydrogen, and oxygen atoms as the starting materials for their super-long chains of carbon atoms. Carothers decided that they had exhausted the possibilities of this family of organic compounds. So he began to investigate the properties of polyamides, a family of organic compounds that included atoms of nitrogen in addition to atoms of carbon, hydrogen, and oxygen.

Over the next several years, the Carothers team investigated the properties of many of these new man-made polymers. Two of these polyamides proved under testing to have the strength, high melting point, and elasticity that were needed to create a useful product that could be sold. In the summer of 1935, Elmer Bolton, one of the chief decision makers at DuPont, had to decide which of the two was going to be the commercial product.

One of the new polymers was made from castor oil, and this was the one that Carothers preferred. Castor oil is a complex organic compound that is derived from the pressing of castor beans. But there would never be enough castor oil in the natural world to make large quantities of the new fiber. To manufacture castor oil artificially in the quantities that would be needed for the production of the new polymer fibers would require years of research and the construction of a new factory.

The other candidate for production was made from chemicals that were readily found in the residue from oil refineries. These starting materials were cheap and abundant. Bolton could not see a time when they would not be available in huge quantities. This new polyamide melted at 263-265 degrees celsius and had high strength and elasticity. So Bolton's choice was an easy one. He chose the polyamide that was made from oil refinery residue.

At this point, Wallace Carothers began distancing himself from his discovery. His genius was in chemistry. Other men were put in charge of actually producing commercially what he had discovered.

Also, the Great Depression, which had begun in 1929 and was still continuing, meant that DuPont had less money to spend on "pure research." So Carothers' staff was cut. From a high of 12 chemists, he now had just three assistants.

While he'd always been subject to fits of depression, even before the time that he had been hired by DuPont, his condition got worse. Then, surprisingly, he got married late in 1936 to a woman named Helen Sweetman. But marriage didn't seem to make Carothers any happier, not even when he learned that Helen was going to have a baby.

Then, early in 1937, there was a crushing blow. His sister died suddenly. He was very close to her and her loss caused him a great deal of pain.

On April 29, 1937, just two days after his 41st birthday, Carothers checked into a hotel in Philadelphia, Pennsylvania. That night, several people heard loud groans coming from his room. Alarmed, they called the front desk and a hotel employee ran to the door, but he was too late. Wallace Carothers lay dead on the floor. He had committed suicide.

After Wallace Hume Carothers and his team created nylon, DuPont manufactured it in great quantities. These bobbins each hold a large length of nylon.

Epilogue

Better Things for Better Living

Not long after Carothers' death, a patent for the new polymer that his team had developed was issued. It was one of more than 50 patents that the company received because of his work.

Other departments at DuPont Company were given the task of constructing the large commercial-sized facility to produce the new polymer. They also had to design and produce the equipment that would be needed. Over 230 chemists were employed on the project.

But before they could produce it commercially, they had to decide on a name. DuPont executives examined more than 400 possible names, such as Novasilk, Tensheer, Terikon and Amidarn. One man even suggested Duparooh, an abbreviation for "DuPont pulls a rabbit out of a hat," a reference to the almost magical way in which the new fabric had been discovered.

In August, 1938, DuPont announced the name of the new fiber. It was called "nylon." But "nylon" was never a trademark. It was a name given to all of the man-made polyamide fibers, like "wood" or "steel."

Its first commercial use was in making the bristles for toothbrushes. But soon afterward, DuPont announced plans that would make nylon much more important in American life. The company would use it to make stockings for women.

The first experimental nylon stockings were knitted in February, 1939. Later in the year, a giant leg wearing a nylon stocking at the San Francisco World's Fair created interest and awareness of the new product.

And nylon stockings caused a sensation when they were shown at the 107-foot-high DuPont Tower of Research at the New York World's Fair in 1940. Above the main desk at the DuPont exhibit were written the words, "Dedicated to the men and women who, through their contributions of labor and enterprise capital to the chemical industry, are creating Better Things For Better Living."

So it wasn't any wonder that many women wanted to buy them when they finally went on public sale for the first time on what became known as "N" Day, May 15, 1940. Over 90% of all nylon production was devoted to women's stockings. In November, 1941, nylon was used to make the uniforms worn at a professional football game in New York's Yankee Stadium to prove its strength and resistance to abrasion.

At this time, all over the United States, chemistry became popular. People began to believe that science could make life for all Americans better and scientists were portrayed as a new kind of national hero. An article in the *National Geographic* magazine from 1939 described the popularity of chemistry sets among young people: "Toy chemical sets, played with by fascinated boys and girls, make smoke, smells, and bubbling magic in a million homes. In countless classrooms other groups stain their clothes and fingers monkeying with test tubes and tiny bottles."

In a short time the membership in the American Chemical Society rose to 22,245 member chemists. They worked not only in the nation's schools but at over 2000 research laboratories across the country.

In December 1941, the Japanese attacked Pearl Harbor and put the United States into World War II. Chemists and all of their laboratories began to work on the war effort. During the war, all of the resources of DuPont and the rest of American industry were devoted to the effort to win the war.

All the nylon that was produced was used by the military. Parachutes were made of the same kind of nylon cloth that had been used in football uniforms because it was moisture- and mildew-resistant. Made into vests, nylon provided protection from the fragments from bursting shells. It was also used for ropes, tents, uniforms, the noses of large bomber airplanes, mosquito netting, even shoelaces. Nylon reinforcements in the tires of heavy bombers allowed them to land on airstrips that were gouged out of rough terrain.

A large American flag made of nylon flew at the White House to demonstrate that the country could do without Japanese silk.

After the war, DuPont introduced a whole family of new man-made fibers, including those that went by the brand names Orlon, Dacron, and Decton. These fibers were used to make fabric for clothes that could be stained with ink, iodine and ketchup, tossed into a washing machine, and drip-dried on a clothes line. They would be clean, dry, and unwrinkled. These clothes were called "wash-and-wear" and "drip-dry." Soon uniforms for the military, police, firemen, and airline stewardesses were made from Dacron.

Over the years, DuPont has made and sold many different kinds of nylon fibers. Today, third- and fourth-generation manmade polymer fibers continue to amaze people with their incredible strength and usefulness. Kevlar, one of these new fibers, is used to make bulletproof clothing and the wings of supersonic stealth jet planes that do not appear on radar screens. It is stronger than steel. Over 10,000 tons of kevlar are produced every year and woven into a magic cloth that not even Wallace Carothers might believe.

And while nylon continues to be used as fibers for hosiery and lingerie, it also has other important uses in solid form, such as ball bearings, and in medicine where it is used to make artificial limbs and sutures. It's used to insulate wires because it's fire-resistant. Mixed with fiberglass, it makes auto body parts.

It's sad that Wallace Hume Carothers never lived to see the success of nylon. He would have been amazed at the world today, full of its manmade fibers and plastics.

Even with his success, he might have been embarrassed to be called a hero. But Wallace was a leader if not a hero. He was one of the men who made people believe that the imagination, not the products of the natural world, is the only limiting factor to what people can accomplish.

Wallace Carothers Chronology

- 1896, born on April 27 in Burlington, Iowa.
- 1901, moves with family to Des Moines, Iowa, where his father becomes an administrator at a business college. In the years that follow, Wallace will spend a lot of his free time doing science experiments.
- 1914, graduates from North High School in Des Moines, Iowa and enters the business college where his father works.
- 1915, enters Tarkio College, where he teaches business and starts to study chemistry.
- 1920, graduates from Tarkio College, after teaching chemistry there for two years. In the fall, he begins graduate work in the chemistry department at the University of Illinois at Urbana.
- 1924, receives his Ph.D. and becomes a full-time assistant at the University of Illinois.
- 1926, goes on summer vacation to France and begins teaching at Harvard University in the fall.
- 1928, goes to work for the DuPont chemical company to do pure research there for the next ten years. Gradually, other researchers will join his team and help him study polymers.
- 1930, discovers early form of nylon.
- 1936, marries Helen Sweetman.
- 1937, dies on April 29 at the age of 41.

Nylon Timeline

ca. 2600 B.C.: Silk is discovered.

ca. 200 A.D.: Japan begins making silk.

1927: The DuPont chemical company decides to hire scientists to do pure research that its leaders hope will one day develop products or inventions that will make profits. Wallace Hume Carothers is the first chemist hired. Gradually, other researchers will join his team.

1930: A member of Carothers' team accidentally discovers how to make synthetic rubber. In the years that follow, Carothers and his team will do research concerning synthetic fibers.

1935: The DuPont company executives decide to start large-scale production of Carothers' important discovery, nylon.

1938: On October 21, the New York Times announces to the public that the DuPont Company is going to make nylon and that it is expected to one day replace silk.

1940: Nylon stockings go on sale to the general public on May 15 and are an immediate sensation.

1941: U.S. enters World War II and all nylon production goes into the war effort.

1945: World War II ends and nylon soon becomes commercially available.

1980: Nylon is sold in the amount of 7.5 billion pounds a year.

For Further Reading

Adams, Simon, *World War II* (New York: DK Publishing, 2000).

Arnold, Nick, *Chemical Chaos* (New York: Scholastic Press, 1998).

Blashfield, Jean F., *Carbon* (N.P.: Raintree/Steck Vaughan, 1998).

Gilchrist, Cherry, *Stories from the Silk Road* (New York: Barefoot Books, 1999).

Richardson, Adele, *Silk* (Mankato, Minnesota: Creative Company, 1999).

Stwertka, Albert, *A Guide to the Elements* (New York: Oxford University Press, 1999).

Glossary of Terms

Black powder: an explosive
Boiling point: the temperature at which a liquid changes into a gas
Carbon: a chemical element which is found in all living things
Chemistry: the science which studies the composition, structure, and properties of substances
Civil War: war fought between the North and the South (1861 to 1865)
Commune: a small community, whose members share common beliefs
Compound: something that is made up of two or more chemical elements
Elder: a church leader
Element: one of the 100 or so of the most simple substances containing only one kind of atom (oxygen is one)
Fluent: able to speak well
Freshmen: students in their first year at a college or university
Loom: the apparatus used to weave thread or yarn into cloth
Lubricants: substances that reduce friction between moving parts
Melting point: the temperature at which a solid changes into a liquid
Molecule: the smallest amount of something that can exist by itself; a molecule is made up of one or more atoms
Mulberry: a tree that has leaves that silkworms eat
Ocean liner: a huge ship, used to transport passengers across the seas
Organic: having to do with life; of, related to, or containing carbon atoms
Polyamide: an organic polymer made up of carbon, oxygen, hydrogen, and nitrogen
Polymer: an extremely long molecule (organic polymers have long chains of carbon atoms)
Serendipity: the making of a discovery by accident, finding something that you were not looking for
Sericulture: the science of growing silkworms and making silk
Shorthand: a system of rapid handwriting which uses abbreviations and is used during dictation or to make notes
Silicon: a nonmetallic element found in the earth and used to create glass, among other things
Still: an apparatus for distilling liquids
Tuition: the money a student pays to attend classes at a college
Undergraduate: a student at a college or university who is working toward a bachelor's degree

Index